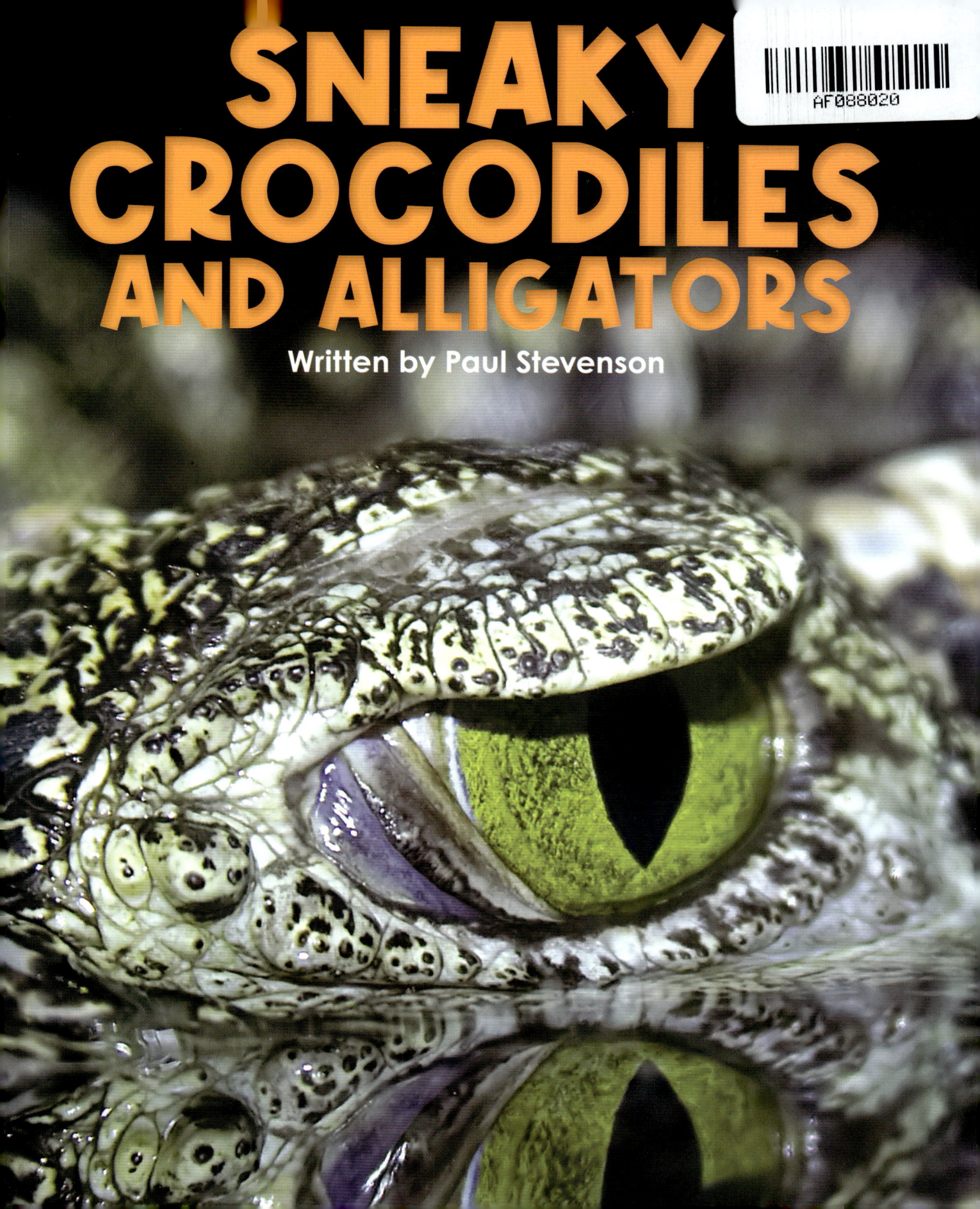

SNEAKY CROCODILES AND ALLIGATORS

Written by Paul Stevenson

CONTENTS

Man-Eater	4
Caught!	6
Monster Crocs	8
Fierce Hunters	10
Silent Attack	12
Croc Goggles	14
Alligator Attack	16
Anti-Attack Tips	18
How to Catch a Croc	20
Ancient Killers	22
Croc Farms	24
Croc Mothers	26
In Danger!	28
Final Facts	30
Glossary	31
Index	32

All words in **BOLD** can be found in the glossary.

First published in 2024 by
Hungry Tomato Ltd
F15, Old Bakery Studios,
Blewetts Wharf, Malpas Road,
Truro, Cornwall,
TR1 1QH, UK.

Copyright © 2024 Hungry Tomato Ltd

No part of this publication may be reproduced, stored in a retrieval system, or transmitted in any form or by any means, electronic, mechanical, photocopying, recording, or otherwise, without prior written permission of the copyright owner.

A CIP catalogue record for this book is available from the British Library.

ISBN 9781916598768
Printed in China

Discover more at
www.hungrytomato.com

Neither the publisher nor the author shall be liable for any bodily harm or damage to property whatsoever that may be caused or sustained as a result of conducting any of the activities featured in this book.

DISCLAIMER:
The people who hunt crocodiles and alligators are experienced professionals. Under no circumstances should you try engaging with a crocodile or alligator yourself, as it is extremely dangerous!

MAN-EATER

For 14 years, there was a **man-eating** crocodile on the loose!

He had killed over 80 people who lived beside Lake Victoria in Uganda, Africa.

THE LOCAL PEOPLE WANTED HIM **DEAD!**

CAUGHT!

Hunters tracked the crocodile for three days and nights. Eventually, they found him. The hunters used ropes to trap him. After a struggle, the monster croc was caught.

The monster croc weighed the same as a small car and was almost 5 metres long.

Park rangers decided to take the man-eater to a crocodile farm. They drove him to the farm in the back of a pick-up truck.

NOW THE 76-YEAR-OLD KILLER CROC IS EATING CHICKEN INSTEAD OF PEOPLE!

MONSTER CROCS

Large, dangerous crocodiles do not just live in rivers and lakes. Some crocs swim far out to sea, too.

Cuban crocodile

Saltwater crocodiles live in and around Southeast Asia and Northern Australia.

Saltwater crocodiles are very dangerous animals. They kill around 1,000 people a year!

The largest crocodile EVER caught and placed in **captivity** was a giant crocodile called Lolong. He was a saltwater crocodile who was 6 metres long and weighed 1,075 kilograms!

He was so big that it took around 100 people to capture him!

Unfortunately, Lolong died in 2013, but his title of "largest crocodile" has not been beaten yet!

Saltwater crocodiles are the largest crocodile species

FIERCE HUNTERS

Crocodiles and alligators are fierce hunters. They usually catch fish and birds. Sometimes they catch land animals, such as antelopes. They catch them when they come to the water's edge to drink.

Crocodile eating an antelope

WATCH OUT!

Large crocodiles also eat people if they are on their **territory**! Humans are food, just like other animals.

Man-eating crocs are becoming more common. This is because more people are living close to the water where the crocodiles hunt.

Signs are usually put up to warn people when crocodiles are nearby

SILENT ATTACK

A crocodile or alligator stays out of sight in the water. It tracks its prey by smell.

The crocodile glides up silently to the prey. Just its eyes and nostrils are above the water. Then, it springs forward in a quick attack.

Prey, like this wildebeest, can be grabbed by the crocodile's jaws and dragged underwater. Crocodiles kill their prey by shaking them or **drowning** them underwater.

Crocodiles can't chew

Their teeth are used for holding and stabbing into prey

Most food is swallowed whole, but larger meals are ripped into chunks

Crocodiles swallow pebbles to help **grind** up food in their stomach.

CROC GOGGLES

Crocodiles have excellent eyesight which helps them spot prey from far away.

Crocodiles can see just as well at night as in the daytime. They can also see well underwater.

They have an extra eyelid that works like goggles.

Eyelid

The layer slides sideways across the eye

See-through layer of eyelid

ALLIGATOR ATTACK

Alligators are smaller than crocodiles but just as dangerous!

An alligator's scales are dark green or black, while a croc's are light green or brown

An alligator's head is shorter and wider than a croc's. Its snout is also rounder and more U-shaped

WATCH OUT!
In Florida there are, on average, 7 reported alligator bites each year!

Most attacks happen when people are swimming or wading in the alligator's water.

Sometimes, people and wild animals come into contact by chance!

ANTI-ATTACK TIPS

REMEMBER THESE RULES:

- Never feed crocodiles. It will make them lose their fear of people.

- Never get closer than a body length away from a crocodile. It can attack you faster than you can get away.

- If a croc opens its mouth or hisses, it is time to leave.

NEVER FORGET, CROCS CAN RUN AS FAST AS HUMANS!

HOW TO CATCH A CROC

Crocodile hunters know a few tricks. They trap crocodiles in nets. They use chunks of meat as bait.

They put rope around the crocodile's mouth. The rope is so strong that the crocodile's jaws can't break it.

A man catching a crocodile that had eaten the bait put in the river

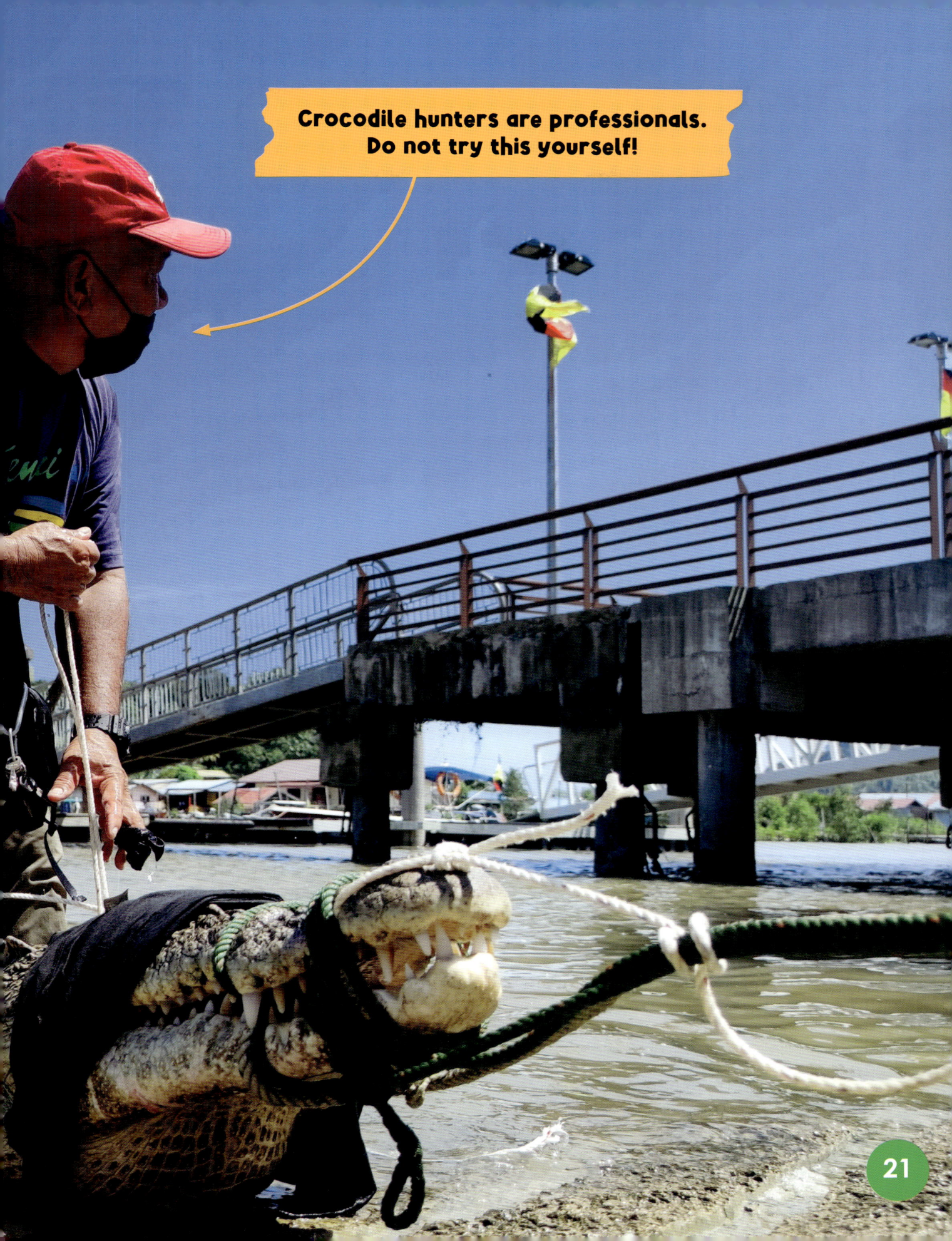

Crocodile hunters are professionals. Do not try this yourself!

ANCIENT KILLERS

Crocodiles have been the top killers in lakes and rivers for over 200 million years!

In that time, no other hunter has **evolved** that can outkill the crocs in water.

This is what a Deinosuchus would look like if it were alive today!

Crocodiles are not dinosaurs, but they are related to them. Crocodiles' ancestors lived at the same time as the dinosaurs!

Crocodiles are **reptiles**. Snakes, lizards and turtles are also reptiles.

The biggest crocodile of all time was Deinosuchus. It lived 83 million years ago, back when dinosaurs existed. At 12 metres long, it would have been longer than a school bus!

CROC FARMS

Sometimes, crocodiles and alligators are kept on farms. These aren't like your usual animal attractions, though!

They are grown, in farms, for their skin and meat. The skin is used to make handbags and shoes. Crocodile meat is very low in fat. Some people eat it instead of beef.

Crocodiles often get hurt in fights. They live in very dirty water, but their cuts don't get **infected**.

Scientists think crocodile blood could therefore be used to make medicines.

Sadly, crocodiles are sometimes bred in overcrowded, small enclosures like this one

CROC MOTHERS

Female crocodiles are very gentle with their babies.

A mother croc buries her eggs in a nest made from earth and grass.

When the babies are ready to hatch, they begin to **chirp**.

When the croc mother hears the chirps, she breaks open the nest. The babies break out of their eggs and slither into the water.

People used to think that crocodiles ate their babies. The crocodile mothers were actually giving their babies a ride inside their huge mouths.

Baby crocodiles riding on their mother's back

IN DANGER!

There are 24 different types of crocodiles and alligators.

Many species of crocodiles and alligators are in danger of **extinction**. Gharials, Chinese alligators and muggers are all **endangered**.

GHARIALS

Conservation status: Critically endangered
Threats: Habitat loss, **unsustainable** fishing practices and hunting.
Identification: These weird-looking crocs have long snouts with a knobbly growth at the end.

CHINESE ALLIGATOR

Conservation status: Critically endangered

Threats: Habitat loss and contaminated food resources.

Identification: It is the only alligator that lives outside of America.

MUGGER

Conservation status: Vulnerable

Threats: Habitat destruction

Identification: This medium-sized crocodile has the broadest snout of all crocodiles!

FINAL FACTS

LARGEST:
Saltwater crocodile up to 6 metres long

SMALLEST:
African dwarf crocodile up to 1.9 metres long

- A large crocodile can stay underwater for up to two hours.

- Crocodiles communicate by roaring and making buzzing noises underwater.

- Birds sometimes pick at food stuck between a croc's teeth. The bird dentists help keep the croc's teeth clean!

GLOSSARY

bait - a piece of food used to attract an animal into a trap.

captivity - living in a cage or enclosure, such as in a zoo or on a farm.

chirp - a high-pitched, squeaky noise made by baby crocodiles.

drowning - dying by breathing in water.

endangered - at risk of becoming extinct. When an animal is endangered, there is a risk that it may die out altogether.

evolved - when an animal changes over a very long time to live in a different way.

extinction - when a type of animal or plant has died out and there are none left.

grind - to crush something into small bits.

infected - to be attacked by germs that cause illness.

man-eating - an animal that eats humans.

park ranger - a person who works in a wildlife park and cares for wild animals.

prey - an animal that is hunted by another animal as food.

reptile - a cold-blooded animal with a backbone. Most reptiles lay eggs. Snakes, lizards, turtles and crocodiles are all reptiles.

saltwater - water that has salt in it, such as seawater.

territory - the area where an animal finds its food and its mate. An animal will fight to protect its territory.

unsustainable - something that's not able to continue for very long.

INDEX

A
African dwarf crocodiles 30
alligators 10, 12, 16-17, 24, 28-29

B
baby crocodiles 26-27

C
captivity 9, 31
catching crocs 20-21
Chinese alligators 29
crocodile attacks 8, 12, 18
crocodile farms 7, 24-25
crocodile hunters 6, 20-21
crocodile mothers 26-27

D
dinosaurs 23

E
eating 13
eggs 26

extinction 28-29, 31
eyes 12, 14-15, 31

G
gharials 28

H
hunting skills 10-11, 12

J
jaws 12, 20

L
length 7, 9, 23

M
man-eating crocs 4, 6, 11, 31
meat 24
medicine 25
monster crocs 6-7, 8-9, 22-23
mouths 18, 21, 27
muggers 29

P
prey 10, 12-13, 31

R
reptiles 23, 31

S
saltwater crocs 8-9, 30-31
skin 24
smell (sense of) 12
snouts 16, 28-29
stomachs 13

T
teeth 13, 30
territory 11, 31

W
weight 6, 9

Picture credits:
(t=top; b=bottom; c=centre; l=left; r=right):
Shutterstock: 1, 14-15, 28, 18-19. Arunee Rodloy 26tl; Chase D'animulls 16, 31b; Emily Barker 11b; Eugeny Popov 13; Herrieynaha 20-21; L-N 2-3; Marc Pletcher 26-27; Pomln0z 9b; Sitthipong Pengjan 24b; TJOGR 25b; Wildestanimal 8t. AFP/ Getty Images: 7. Bjorn Svensson/ Science Photo Library: 6c. China Photos/ Reuters/ Corbis: 29t. Christian Darkin/ Science Photo Library: 22/23. Frinz Polking/FLPA:10. Joanna Van Gruisen/ Ardea: 29b. Juniors Bildarchiv/ Fotosearch: 30b. Mike Parry/ Minden Pictures/ FLPA: 8bl. NHPA/ Eric Soder: 17b. Stephen Crawford/ Alamy: 25t. Superstock/ age fotostock: 4-5, 12. W. Perry Conway/ Corbis: 18-19.

Every effort has been made to trace the copyright holders, and we apologise in advance for any unintentional omissions. We would be pleased to insert the appropriate acknowledgements in any subsequent edition of this publication.